Guy de Maxence AFANDA

$$E = mc^2$$

or

$$m^3 = Eq^2$$

INTRODUCTION

The difference between the mathematician and the physicist appears clearly in front of the length. The mathematician looks for the measure of the length; the physicist seeks its definition.

Physics is the discourse on nature. "Nature", that is the permanently observable, the permanently existing.

Man therefore cannot evade the common characteristics or common properties that inexorably gather him with the rest of the universe.

Also, any panoramic attention to the global, advocating the presence of the general characteristics or general properties attached to everything that exists.

Physics, or speech on the permanent is its ultimate mission, the discourse on permanent features of the existence (matter, motion, energy, strength ...). Its evolution is necessary to go through several stages of human thought.

There was first the theological physics, in the style of biblical genesis. Here, the principles are the observation, the divine will and the analogy (the thunder is the scolding of God, the storm is the wrath of God, the rain is watering of God, phenomena are acts of God otherwise versions of Acts of God, the solar system, for example, is a clock of God as well as the alternation of day and night is a decision of God).

4

Then there was the mythological physics. The myth is an arguing narration on experience or the real. But it is a priori story. Mythological physics is constructive or presumptive, based on observing, interpretation and comparison. In the manner of Aristotelian physics, it describes the phenomena by deporting them mentally from their objective scene, to rally them to the persuasions, to rebuild them according to intellectual frames. For example, according to Aristotle, things fall because they are loaded or besieged by a drooping constituent. Still, "the rest is the natural state of things".

So far, the physicist speaks of nature at the place of nature itself. He replaces it for the content of the utterance. The observation is not involved, but the observation data pronunciation. For example, the falling object shows the movement of a thing moved without pulse, or more precisely without motor. The Aristotelian statement draws on the ingenuity to set. Even the Newtonian and Einsteinian contributions do not meet permanently.

Still, when mathematical physics emerges, it occurs as physics based on observation, experimentation and mathematical formulation of phenomena. The experiment was so successful under the copy phenomena and observation data, and to point to the principles and causes which determine the phenomena. Moreover, it is unlikely to obtain otherwise faithful mathematical certainties information, without referring directly to the phenomena themselves.

Today at least, if a visitor enters the world of fashionable physics in the world, he will find present or ancient actors, results in force or obsolete, stories indicating overactive brains and devices, and in the background, the history of physics. This story is a string prowess but a bundle of discoveries, rebuttals, consensus and controversy. It is precisely to silence the controversy that the great masters of this universe appeared. Galileo and his contemporaries (Descartes, Kepler ...) and their heirs (Newton, Coulomb ...) arrive and end the ancient Aristotelian physics setting out measurable theses (giving the ability to measure phenomena).

Today, however, reefs require the abandonment of devoted certainties for the advancement if not the coherence of mathematical physics. Indeed, how to reconcile the Newtonian celestial mechanics and the theory of general relativity, while they are both strong but tight? How is it possible that the generality of the general theory of relativity is effectively hindered by quantum mechanics? And the unit theory that is struggling to emerge as the big bang theory states or establishes the unity of the universe.

The work presented in the following pages, was motivated against these pitfalls. To avoid exhausting and fruitless labor if not counterproductive labor, it was necessary first, obey the adage, "when we got lost along the way, you better go back to the place of departure, leaving with insurance"; secondly, it was necessary to check how far the assertion of Gaston Bachelard is true, that the current physics is a factory of phenomena, to remove the clutter of the made up assertion, if not the pollution of the conjectural affirmation, harmful to the truth in experimental physics. It was therefore necessary to reconcile out classical mechanics, the theory of relativity and quantum mechanics, so that mathematical physics forms a block or a one-frame body of complementary theories, abandoning necessarily the celestial mechanics of Newton, the Einstein equations and the ideas underway in quantum mechanics. It must be confusing positively (or observably) the weight and inertia, expand diffraction equations and interference to all mobile and mobile of all kinds (wave particle, body systems ...), and confuse the Planck universal constant **h** with a moment.

CHAPTER 1: **The duality theory**

Does the light really have a speed limit? In other words, can it not go faster than the c known?

1) The optical geometry indicates that the index of refraction of a medium is $n = \dfrac{c}{c'}$, where c' is the speed of light in the most refractive medium. In other words, the zero refractive index gives possible infinite speed. Observation shows that in fact the speed of light in a medium always appears as the limit speed of light in that medium. Why would it not be the case for any mobile running in any medium? In other words, mobile must be as light, that is to say have a variable speed limit in every medium traversed.

2) Let now S and S' be two Galilean reference systems in the same medium. When S' is moved in S with the velocity u, S' and S being each of three orthogonal axes, any direct path in S' appears in S under the aspect:

$$\begin{cases} dx = dx' + u_x dt \\ dy = dy' + u_y dt \\ dz = dz' + u_z dt \end{cases}$$

, and in S' in the aspect:

$$\begin{cases} dx' = dx + u_x dt' \\ dy' = dy + u_y dt' \\ dz' = dz + u_z dt' \end{cases}$$

It thus appears that:

1 °) $u^2 dt^2 = u'^2 dt'^2$; $(v-u)^2 dt^2 = v'^2 dt'^2$ then:

$u'^2 v'^2 = v^2 (v-u)^2$; now: $v = v' + u - u(\dfrac{dt - dt'}{dt})$;

and therefore: $u'^2 (v-v')^2 = u^4$; and then to the outside observer to S': $(v-v')^2 (v-u)^2 = u^2 v'^2$; in the case of the mobile in S' moving in the same direction and the same sense that S' itself, observe:

$v^2 - v(v' + u) + 2uv' = 0,$ be here:

$$v = \frac{v' + u + \sqrt{(v' + u)^2 - 8uv'}}{2},$$

if: $(v' + u)^2 > 8uv'$; and: $v = v' + u$, if: $(v' + u)^2 \le 8uv'$.

2) The length of an object does not necessarily change with the movement of the reference system; Indeed, like the line segment [A, B]; its position is in S:

[A,B]$_S$ = (x$_B$ - x$_A$; y$_B$ - y$_A$; z$_B$ – z$_A$) ; but: x$_B$ = x$_B$'+ a, and x$_A$ = x$_A$' + a, then: x$_B$ - x$_A$ = x$_B$'- x$_A$' ; and therefore:

[A, B]$_S$ = [A, B]$_{S'}$.

Now, taking into account the movement of S' according to the conditions already indicated, we note that:

$dx_B - dx_A = dx_B' + u_x dt - dx_A' - u_x dt = dx_B' - dx_A'$; and
similarly: $dy_B - dy_A = dy_B' - dy_A'$, $dz_B - dz_A = dz_B' - dz_A'$; so
that if [A, B] is solid, firm, or plain:

$$d(x_B - x_A) = d(y_B - y_A) = d(z_B - z_A) = 0 .$$

3) Like all things, time speed and length are combinations of absolute and relative

4) All Galilean reference locations (respectively all reference systems) are independent for observation but equivalent for the definition and formulation of phenomena.

In dynamics, we have vector relation (the vectors are in bold):

$$\textbf{p'} = \textbf{p} - m\,\textbf{u} = \textbf{p} + m'\,\textbf{u'}$$

*

The theory of general relativity states and is based on the equivalence of acceleration and gravity. Indeed, if S' is subjected to a pulse, the external observer sees the inside observer drop in the opposite direction to that of the pulse. For indoor observer there is a birth of weight.

11

Canonically taken, within a reference system subjected to a pulse, there is equivalence of the acceleration and gravity, out of this system, the acceleration and gravity are simultaneous, equal and opposite.

The new composition of forces to the outside observer must come by extension to the dynamic of relations:

$$\begin{cases} d^2x = d^2x' + du_x dt \\ d^2y = d^2y' + du_y dt \\ d^2z = d^2z' + du_z dt \end{cases} \text{; and} \qquad\qquad \text{therefore:}$$

$$\begin{cases} d^2x' = d^2x + du_x dt' = d^2x + g_x dt^2 \\ d^2y' = d^2y + du_y dt' = d^2y + g_y dt^2 \\ d^2z' = d^2z + du_z dt' = d^2z + g_z dt^2 \end{cases}$$

They retain the remarks leading to the theory of relativity reformulated, whose occurrence in that: all reference locations (respectively all reference systems) are independent for observation but equivalent for the definition and formulation of events; this is the general principle of relativity reformulated here.

In dynamics, this gives: $d\mathbf{p}' = d\mathbf{p} + \mathbf{P}dt = d\mathbf{p} + d(m'\mathbf{u}')$, $\mathbf{P} = m\mathbf{g} = - mdu/dt$, is the relativistic weight for the inside observer.

Overall, $\mathbf{g} = -d\mathbf{v}/dt = GmN/S$, where S is the pressed surface; G is now the dynamic moment of the pulse on

the body mass m, manifested by vibration, oscillation; indeed, ρ is the density of the body, $G\rho = \Omega_\omega$
 Recall that:

$$\Omega_\omega = \frac{a''}{a} + \frac{1}{a}\left(\frac{da\omega}{dt} + a'\omega\right)i - \omega^2$$

That said, **g** is oriented in the same direction as the pressed surface, and because of his equality with the acceleration **(g =- γ)**, has the same intensity everywhere; so that: $\Sigma\ \mathbf{F}_{ext} = -\ \Sigma\ \mathbf{P}_{created}$. And if m_j is an m element, the weight of m_j is:

$$\mathbf{P}_j = m_j\ \mathbf{g}_j\ \mathbf{N} = Gm_j m\ /\ S.$$

The immediate consequence for celestial mechanics is that it must be abandoned the Newtonian celestial mechanics. Otherwise, it should be recognized according to the theory of Newton, when the sun is in the east things fall to the west, when at the zenith things fall down, when in the west things fall to the east, and when at the nadir things fall up.

CHAPTER 2: revisited quantum mechanics

To discourse on quantum mechanics, a preliminary dual is obligatory:

1) The general principle of relativity must remain valid; for a quantum observer, quantum event is observed with respect to a reference system; this event has the same definition as in any other location or reference system, but can not necessarily be perceived in the same way everywhere

2) The quantum phenomenon is a member of a series; in fact, cutting a line L into n equal parts l, if it is continuous it is written as: $L = nl$, and if it is discontinuous, it can be written: $L_n = nl + (n-1) i$, where i is the constant interval between plots segments.

Thus, in general, if we write: $L = \Sigma l_j$, then we can write:

$$L_n = \Sigma l_j + \Sigma i_{j-1} , \text{ with: } j \geq 1.$$

In quantum mechanics, this is to design an energy series: $E_n = \Sigma E_f + \Sigma E_c$, where E_f is the formation energy, and E_c emission energy that remains as the kinetic energy of quanta.

Further, it should be noted that the quantization of a quantity X is: $\dfrac{dX}{df} = \dfrac{dX}{vdn} = \dfrac{x}{v}$, where f is the frequency of formation, the transmission frequency v, n the number of formed quanta and x the quantum produced.

In the case of quantum mechanics, X is quantized energy as:

$$\frac{dE}{df} = \frac{e}{v} = h$$

.

CHAPTER 3: The general diffraction

A- **DIFFRACTION WITH ONLY ONE OBSTACLE**

Physics holds that the diffraction is the deflection undergone by the wave propagation (acoustic, light, radio, x-rays ...) when it encounters an obstacle or passes through a hole equal to their wavelength dimensions.

Take any wave in observation:

1) The wave is the movement of a thing whose shape changes

2) The waving is the displacement of the wave.

So: the wave is something moving; no wave at rest.

Wave mechanics can then be coated or continuous aspect or discontinuous aspect.

Wave mechanics in continuous aspect concerns the thing whose shape changes. Besides, there is the discontinuous wave mechanics whose theme is the wave-particle appearance. This has two forms: the waving corpuscle and the association: wave + corpuscle.

In the case of the particle undulating or corpuscular wave, the quantum takes the form:

$\int \hbar\omega \partial s = Gm^2 = 0$; where G is the dynamic moment of the pulse on the corpuscle.

In the case of: wave + corpuscle, the wave and the particle are mutually stationary, because associated in the same run. Thus, there is conservation of mechanical energy, such as: $\hbar\omega = mv^2$; with: $\hbar\omega = hv_\lambda / \lambda$, where v_λ is the full speed of the wave; equality is:

$mv^2 = hv_\lambda / \lambda$. But precisely here: $v_\lambda = v$, and thus: $\lambda = h / mv$. The meaning of this formula is that it is only possible if the particle that moves is associated with an independent wave. Otherwise, more accurate work on this principle has the relationship: $\Delta(\hbar\omega) = \Delta (mv^2)$. Thus the kinematics characteristic relationship is: $s = n\lambda$, or more precisely: $s = \Sigma\lambda_{i,}$, where s is the path of the particle and the other is the path of the wave.

But in the other case, the ripple and translation are concomitant, since the particle may wave at the same position. The specific formula for this case is:

$\int\hbar\omega\partial s = nhv = Gm^2$. Here, the length of the wave is carried by an independent length, such that:

$$ds = d\Sigma\lambda_{i} = vdt.$$

In short, in a case there is a combination of wave and a translation, and in the other, there is a combination of two translations.

To extend these results to non-material waves, just replace mv with appropriate entries. And consequently, it is each time, to identify a wave propagating in a wave race. The race is the size: $\int pdt$, noted Γ. Now at first experience, it is that: When running object encounters an obstacle, it is deflected in whole or in pieces according to the direction of arrival. The direction of incident stroke being θ or i, diffracted race is:

$\Gamma' = \alpha \Gamma + \beta N$, where βN is the diffracting race ordered to the running object by the reaction of the obstacle. The Crack of the fireball or not is not at issue, because a meteor necessarily undergoes deformation at the shock, simply because of the general principle of diffraction: a run is deflected under equality of action and reaction, at the contact with the obstacle.

So overall, use the formula: $\Delta\Gamma = \Gamma - \Gamma_0$, where Γ_0 is the forward stroke. When the race passes through a thin open hole, then:

$\Gamma_0 = \Sigma \Gamma_{i+1} = \Sigma \Gamma_{2k-1}$, with: $0 \le i \le 2k-1$, as: $\Delta \Gamma = 0$, the open hole being thin in this case the void; k is the relative index of fragility of the mobile incident; let $k = n_c \, d_{o/i}$ n_c is the number of shocks, and $d_{o/i}$ is the density of the obstacle relative to the arriving movable.

B- DIFFRACTION ON MANY OBSTACLES AT THE SAME TIME

Here on the obstacles, the meteor acts as a united field and not as a singularity. This remark is also valid in case there is only a meteor which the contact surface can be flat or varied. In this case, when the runner is simple it easily acts as a singularity, in difference with the cases where it is diverse. In addition, when for example the fireball contact surface is varied and then varies the shock, the above general formula does not apply easily.

That said, a transition with a brief overview on the physical field, present the physical field as the juxtaposition or a troop of quantities of the same kind. It is varied when they are not equal and uniform in the other case. It is united when set on a singularity seat and sparse when set on several singularities at a time.

It is easy to show a united field is uniform necessarily, but not the reverse, since we can write it in the guise of a multiplication. Indeed, one united thing Q can be written as a sum. When combined with another field, multiplication is obtained which makes Q multiplier.

When the field is sparse, writing is not suitable as: $Q = \Sigma q_i$, but: $Q = \}q_i\{_{1 \le i \le n}$. In the specific case of the fireball that hits several obstacles, there is $n\Gamma$ at the shock, where n is the number of obstacles. The formula is used for the meteor with varied impact surface, when it hit a single obstacle. So with: $\Delta(n\Gamma) = \Gamma \Delta n + n\Delta\Gamma + \Delta\Gamma \Delta n$; in stillness we have : $\Delta n = 0$; and when there are holes, Δn is the decline in the field caused by the presence of holes.

C- DIFFRACTION AND REFRACTION

There is refraction when there is a double diffraction under equality of action and reaction. Indeed, with the angle of incidence i, δ the angle or direction of diffraction and r the angle of refraction:

1) Or refraction is just another diffraction

2) Or diffracted run is held in the form of two different runs, the diffracted run and the refracted run

3) Or refraction is codiffraction under equality of action and reaction.

The fundamental relationship of diffraction gives:

$(\Delta\Gamma)_\delta = \beta N$ (deflection caused by the reaction); now: $\beta N = - (- \beta N) = - (\Delta\Gamma)_r$ where r is the angle or direction of refraction $(\Delta\Gamma)_r$ is the refraction (bending caused by the action). Clearly, the refraction is codiffraction. Therefore:

$$(\Delta\Gamma)_\delta + (\Delta\Gamma)_r = 0.$$

Calculations show that the buoyancy is only a special case of the double diffraction, and shape of the first diffraction in general, the following formulation: **a fireball entering or striking any place is pushed or deflected according to the density of the place and according to the depth of penetration or impact.** In this case, at the contact, the obstacle receives impel through action: $F = N\Delta E_{ci}/\varepsilon = \mu\gamma$, where E_{ci} is the incident kinetic energy, ε depth of impact, μ the moving mass of the obstacle and γ the acceleration received by μ. But μ is equal to the density ρ of the obstacle multiplied by the displaced volume V_ε. The thrust of the obstacle or its reaction is: $R = - N\Delta E_{ci}/\varepsilon = -\rho V_\varepsilon \gamma = \rho V_\varepsilon \, g_\varepsilon$, where g_ε is the relativistic gravitation of the moving mass of the obstacle.

D- <u>DIFFRACTION AND OPTICS</u>

When the obstacle is curved or when the impact surface is curved, there is divergence or convergence relative to the center of curvature C. Thus, by designating the line or axis of curvature as the line which has the radius of

24

primary curvature, the study establishes that in the specific case of the uniform curvature, we note that:

1) The directions of all normal pass by C

2) When the obstacle is convex with respect to the incident run, diffracted run meets bending line if the direction of the incident run meets it in the concavity area

3) When the obstacle is concave with respect to the incident run, diffracted run meets bending line if the direction of incident stroke does not meet it at the main radius of curvature.

Optical:

1) The obstacle is the open for ligth or optical medium

2) The bending line is the optical axis

3) Then, the optical event is only possible if at least, in the optical medium, the incident light is in part or in whole diffracted or refracted towards the optical axis.

E- **CONCLUSION: DIFFRACTION AND INTERFERENCE**

25

The two phenomena are related to the contact. Indeed, two things are heading toward each other; at contact, it happens:

1) $\Delta\Gamma_1 - \beta_2\, N_2 = -\Delta\Gamma_2 + \beta_1\, N_1$. And therefore:

$\Sigma(\Delta\Gamma_i) = \Delta(\Sigma\Gamma_i) = \Sigma(\beta_i\, N_i)$, if there is diffraction

2) $\Delta\Gamma_1 - \beta_2\, N_2 = \Delta\Gamma_2 - \beta_1\, N_1$, if there is interference.

Overall, the case is:

$\Sigma(\Delta\Gamma_i + \beta_i N_i = \Delta\Gamma_j + \beta_j\, N_j)$, with: $1 \leq i \neq j \leq n$.

In the case of the wave, it is necessary to link the shock and wave phenomenon by the following formulas:

$Gm = A\,dv\,/\,dt$; $G\rho = \Omega_\omega$; $2\varepsilon dv\,/\,dt = -v^2$; where G is characterized by disruption or shock inflicted on an object of mass m and whose density is ρ; On the impact surface, ε depth of impact, ω the angular frequency imposed on the obstacle, v incident speed.

CHAPTER 4: matter, energy, light

A- <u>MATTER</u>

Matter is no vacuum. And different of vacuum which is essentially something holding nothing or holding space (given as the appearance of nothing); it is characterized by a variety of mechanical chemical or physical states.

1°) The fundamental physical states

There is solid state, which is characterized by molecules strongly fixed on each other. There is the liquid state, which is characterized by molecules which slide on each other. And there is the gaseous state, which is characterized by molecules untied of each other.

However, to understand the intermolecular bonding, use the volume heat, the amount of heat per unit volume. If we denote it Q_v, adhesion or the rate of embrace sought is :

$$\beta = 1 / Q_v = V / Q = 1 / p_c$$

where p_c is the molecular expansiveness. The expansion is in the formula in the form:

$$\beta / Q + d\beta / dQ = dV / Q \, dQ .$$

G with the agitation factor of molecules and ρ the density of the object material body, we get the formulas: $G\beta = (1 / \rho s)^2$ and

$G / \beta = (dv / dt)^2$, where v is the speed of the mobility of molecules. It follows the equation of motion: $\rho s \beta dv / dt = 1.$

2 °) The mechanical states

The most common mechanical state is running. Then there are electricity, magnetism, light and life.

Indeed, the electron is the subject of a characteristic mobility given by the formula:

$$m_e = ev = (ke^2 \theta)^{1/3} = (Ge^3 \omega)^{1/2} = (ke\theta / v)^{1/2} = Ge^2 \omega / v,$$

where v is the velocity of the charge e, k Boltzmann constant, θ room temperature, G the factor of oscillation and which complicates the electron as a biparticule in instability, ω the angular velocity of the electron.

Another consequence of this formula is the formula: $m_e^3 = 2E_c e^2$, for the electron. And in comparison with the chemical structure of the body, it is easy if trying to get picked up, $m^3 = Eq^2$, where m is the mass of the body, the total latent energy E and q latent sum of electrical charges. Analyzing, it is apparent that: $m^3 = qR_\alpha F_c$, where F_c is the strong interaction force and R_α is the radioactivity of the body.

The run of the electron is: $\int m_e \, ds$. In its interaction with the outside it has on the objects, the force: $F_e = dm_e v / dt$.

When the object is rather a different electrical charge, the phenomenon that is born has the form:

$q_1 v_1 + q_2 v_2 = (q_1 + q_2)(v_1 + v_2)$. Therefore:

$q_1 v_2 = - q_2 v_1$. In other words, the charges with the same signs repel each other and the charges of opposite signs attract each other. With N charges, the phenomenon has the form:

$$\Sigma q_i v_j = 0, \text{ with } 1 \le i \ne j \le n .$$

Magnetism is already in power, including that of Negaton where the central force: $F_n = m_e v\omega \, n$, takes the shape of the magnetic force F_μ .

B- **ENERGY**

The energy in physics is the ability to produce the work, according to the formula: $\Delta E = + W$. The initial energy is incident energy equal to the resistance of the subject where energy is exerted. Thus, if (1) is on (2), the energy exerted is: $E_{1/2} = R_2 + W_1$, where W_1 is the work of (1) and R_2 the resistance of (2). Because of the equality of action and reaction, the action-reaction pair form a single couple, and its total energy is constant in the form:

$$E_{1/2} - W_1 + R_1 = E_{2/1} - W_2 + R_2$$

When the system is complex, the total energy is conserved in the form:

$$\Sigma\ (E_{i/j} - W_i + R_i = E_{j/i} - W_j + R_j)\ ;\ \text{with: } 1 \leq i \neq j \leq n.$$

C- LIGHT

For the ontological view, we must return to writing: $m_e\tau$, which indicates that the electron is spread on the tangent of its trajectory. It is therefore in the first idea, a filament. And when the tangent turns, the electron takes the form of a whorlet.

Now in the same way that the electron is associated with the phenomena of attraction or repulsion, when he meets other sources of attraction or repulsion, it undergoes mechanical changes that could lead it to a dazzling stage: radiation. All cases correspond to the variation of the mass of the electron as:

$\Delta m_e = \psi$ (v, θ, Gf), where ψ indicates and is the dependency link of Δv, $\Delta\theta$, $\Delta(Gf)$ and Gf the electric trance. But considering the equivalence of work and heat, it is asking the rate of change: $\Delta m_e / \Delta Q$, where Q is the incoming heat such as when it goes to infinity and temperature with it, it comes to the relationship: $m_v / nhv = ec / nhv + 1 / nc^2$, so : $m_v = ec + hv / c^2$, the mass of the formed quantum, where hv is seen as the emission energy that continues as the resonance of the radioactivity of the radiating source in the receiver or environment. This medium receives an amplitude of vibration such that: $hv - \rho a^3 c^2 = 0$ and $\omega = 2\pi v$. So the thesis that: the propagation of light is the transmission of the vibration of the radiating source to the ambient environment. The proof for this is the disappearance of the light from the radioactivity of the source stops when you turn off a light bulb for example. The light is a transfer of vibration that makes photon a wavelet (a particle-wave), the fruit of waveletization suffered by the environment; the light moves waveletly way.

This corpuscular character command to write that all electromagnetic radiation has the formula:

$$R_n = nm_vc^2 = nec^3 + nh\nu$$; the formation energy

is: $E_f = \sqrt{h^2\nu^2 + k^2\theta^2} - h\nu$; the distance between two

successive quanta δ such that:

$$\frac{k^2\theta^2}{hc}\delta = \frac{hc}{\lambda}$$; so: $k\theta\sqrt{\lambda\delta} = hc$.

The kinematics of the photon depends on that the speed of light must be variable. Otherwise the double diffraction phenomena (refraction and reflection) would not be possible with light. For, to change direction, the meteor must first stop. Thus, the photoelectric effect, indicate that at the shock, the quantum process that arises results by link to change determined by the following law of shocks: $\begin{cases} \Sigma\Delta p_i = 0 \\ \Sigma\Delta E_i = 0 \end{cases}$, such that when the radiation enters an object Ω : $\Delta m_vc^2 + \Delta E_\Omega = 0$.

And then: $0 - m_vc^2 + E_f + E_v = 0$.

$$E_f = \sqrt{E_v^2 + k^2\theta^2} - E_v$$; E_v is a kinetic energy.

CONCLUSION : The uncertainty relations

The measure ē of a quantity is the rate of its presence on the measuring instrument. This can be gained with a margin of inaccuracy Δe = ē-e , where e is the exact measure of the greatness E.

However, two sizes are compatible on the measuring instrument, when concurrent on the measuring instrument. In other words, is to measure the EA sizes and A + E. They are compatible if:

Δae = $\overline{ae}$ – ae ; and: Δ (a + e) = $\overline{a+e}$ – (a+e). If they are incompatible: Δae = ēā-ae ; and: Δ (a + e) = ā+ē – a-e.

Therefore, in the case of incompatibility,

Δ (a + e) = Δa+Δe, and: = Δae =eΔa + aΔe + ΔaΔe.

This is necessarily different from the case of compatibility when the following entries are required:

1°) $\bar{ae}$ = ae + ΔaΔe ; so : Δae = ΔaΔe

2°) $a \mp e$ = a+e + $\dfrac{(a + e)\Delta a \Delta e}{a\Delta e + e\Delta a}$; therefore :

Δ (a + e) = $\dfrac{(a + e)\Delta a \Delta e}{a\Delta e + e\Delta a}$.

Inaccuracies of measures are due to the narrowness of the measuring instrument, to the awkwardness of the measurer, to the vice of measuring instrument, but also in contact with the phenomenon. When the measured and the measuring meet, there is contact such as they influence each other. This gives Δa = Gx and Δe = G'x ', where G and G' indicate the respective suffered disruptions. For a period Δt, accuracy remains inaccessible as: Δa = aωΔt / 2π, for example, where ω is an angular frequency. Upstream and downstream, eddies may fade at the end of this period, but the

37

irreducible uncertainties remain possible if there is a lag between the time Δτ of availability of the greatness, and Δt. Here we must then count:

Δτ - Δt = h / 2W - Δx / fx ; there is certainly possible measure only if: Δτ - Δt ≥ 0.

by

www.ingramcontent.com/pod-product-compliance
Lightning Source LLC
Chambersburg PA
CBHW072258200526

45168CB00016B/2152